Bruno Duarte
Elias de Almeida Silva
Jhonatan Peres de Sousa

BEARING FAULT IDENTIFICATION TECHNIQUES

Bruno Duarte
Elias de Almeida Silva
Jhonatan Peres de Sousa

BEARING FAULT IDENTIFICATION TECHNIQUES

STUDY OF FAULT IDENTIFICATION TECHNIQUES IN ROLLING BEARINGS

ScienciaScripts

Imprint

Any brand names and product names mentioned in this book are subject to trademark, brand or patent protection and are trademarks or registered trademarks of their respective holders. The use of brand names, product names, common names, trade names, product descriptions etc. even without a particular marking in this work is in no way to be construed to mean that such names may be regarded as unrestricted in respect of trademark and brand protection legislation and could thus be used by anyone.

Cover image: www.ingimage.com

This book is a translation from the original published under ISBN 978-620-6-76058-0.

Publisher:
Sciencia Scripts
is a trademark of
Dodo Books Indian Ocean Ltd. and OmniScriptum S.R.L publishing group

120 High Road, East Finchley, London, N2 9ED, United Kingdom
Str. Armeneasca 28/1, office 1, Chisinau MD-2012, Republic of Moldova, Europe
Printed at: see last page
ISBN: 978-620-7-74383-4

SUMMARY

This work aims to present a theoretical reference on some conventional and unconventional techniques when it comes to bearing maintenance based on materials science and machine elements. The study of these techniques was carried out with an emphasis on the main process parameters, such as: lubrication, ultrasound, vibrations, signal processing, mechanical properties and the microstructure of cast iron. Bearing in mind the constant advances for application in rotating machinery.

Keywords: Vibration, rolling bearings, ultrasound, lubrication, materials science, cast iron, mechanical properties, microstructure, signal processing.

SUMMARY

1. INTRODUCTION

For years, the term failure analysis has had a specific meaning in connection with fracture mechanics and corrosion failure analysis activities carried out by static production equipment inspection groups. Often, failures in mechanical equipment reveal a chain of causes and effects (FRED and BLOCH, 2016).

To reduce friction between moving and fixed parts, various types of bearings are used, millions of which we find around us, and it is difficult to assess how much bearings have contributed to our comfort and well-being today. Bearings are found in many household utensils available to housewives, in power tools, lawnmowers and snow blowers, in children's toys, in the air conditioner that cools the house in summer and the oil heater that warms it in winter, and in cars. In industry, bearings are present in almost all types of machinery; transport, in all its forms from roller skates and bicycles to giant sea and land wheels, relies on bearings to speed up and facilitate operations. In this context, rolling bearings are likely to suffer wear because they are subjected to the stresses of the system acting on them. Predictive maintenance is very important in this respect, as it enables the actual condition to be monitored and faults to be identified.

(Francklin, 2009) mentions that nodular cast iron is an alloy composed basically of carbon and silicon, with free carbon (graphite) in the metallic matrix, but in a spheroidal form and began to be used industrially from the 70s onwards, expanding the field of application of cast irons, due to the combination of properties such as high toughness, tensile strength, ductility, resistance to wear and fatigue, making it a competitive engineering material, combining properties previously found only in steels.

The aim of this work is therefore to study the techniques used to analyse bearing failures and, from the point of view of materials science, to report on the relationships that exist in the structure of cast irons and correlate their mechanical properties.

2. OBJECTIVES

2.1. GENERAL OBJECTIVE

To study the techniques used to identify bearing faults in rotating machinery.

2.2. SPECIFIC OBJECTIVES

• To study the efficiency of ultrasound and vibration analysis techniques applied to faults in rolling bearings;

• Investigate the consequences of bearing failures in rotating machinery;

• Study the failure mode by metallurgical analysis.

3. BACKGROUND

According to (Ribeiro 2007, p.3) in the case of rolling bearings, understanding the wear mechanisms and the influence of defects of different sizes on the dynamic behaviour of the machine is the principle of developing a predictive technique that makes it possible to anticipate the most common defects with a safe margin.Planning corrective maintenance interventions relatively far in advance provides better use of available resources, reduces machine downtime, minimises the incidence of environmental impacts and the exposure of maintenance personnel to the risks inherent in maintenance tasks.

4. REVIEW BIBLIOGRAPHY

4.1. INITIAL CONSIDERATIONS

According to (Geitner and Block, 2013) in its simplest definition, a failure can be understood as any alteration to an element or component of the equipment that renders it incapable of satisfactorily performing the function for which it was designed. Common stages preceding the final failure are "deterioration" and "damage", which decrease the reliability and/or operational safety of the equipment or individual components during their remaining useful life. According to (Geitner and Block, 2013) they can be classified as a cause of failure:

1. Failures by design;

2. Material defects;

3. Deficiencies in manufacturing processes;

4. Assembly or installation errors;

5. Inadequate operating conditions;

6. Inadequate maintenance;

7. Inadequate operations

(Vasconcelos and Ribeiro, 2014) They also emphasise:

8. Material fatigue. Poor lubrication.

9. Contamination - inefficient sealing.

10. Improper assembly - improper adjustment

11. Improper handling.

It is essential to plan the failure analysis before starting the investigation. A lot of time and effort can be wasted if the history of the fault and the study of the general characteristics of the problem are not treated with due attention (Ludywig, 1972).

Just like other components in a piece of equipment, when bearings are defective they have characteristic frequencies, depending on the location of the defect. Bearing faults can be predicted by measuring vibration and monitoring the presence of fault frequencies and their multiples, so diagnosis should not take into account amplitude alone. Fault frequencies are calculated taking into account the characteristics constructive of bearings(VASCONCELOS and RIBEIRO, 2014).Because they are used in critical processes, a bearing failure can cause serious damage, either to the equipment itself or due to a production stoppage. Monitoring, analysing and correcting bearing problems are therefore critical operations for which maintenance is responsible.

4.2. ROLLING BEARING

According to (Vasconcelos and Ribeiro, 2014), bearings are one of the most important machine components and one of the most used as a connecting element between components with relative rotational movements. They are present in a wide variety of applications, rotations, sizes and environments. Their failure usually leads to equipment downtime, causing enormous damage to the operation. A bearing is a support for shafts, usually made of cast iron or steel, and formerly also of wood, on which a rotating, sliding or oscillating shaft rests. Bearings are made up of sturdy structures called bases and are generally made of iron or steel. They are also used to dampen vibrations in a given system (Yunus and Cimbala, 2008).According to (Norton, 2000), in order to reduce friction, in addition to lubrication, intermediate mechanisms are used between the shaft and the workpiece, such as bearings, which are rotating elements in machines.

Rolling bearings: These are bearings that contain balls or rollers on which the shaft rests. When the shaft rotates, the balls or rollers also rotate inside the bearing. For example, if balls or rollers are inserted between a shaft and a block, as shown in Figure 4.1 below, the shaft will roll on the balls or rollers.

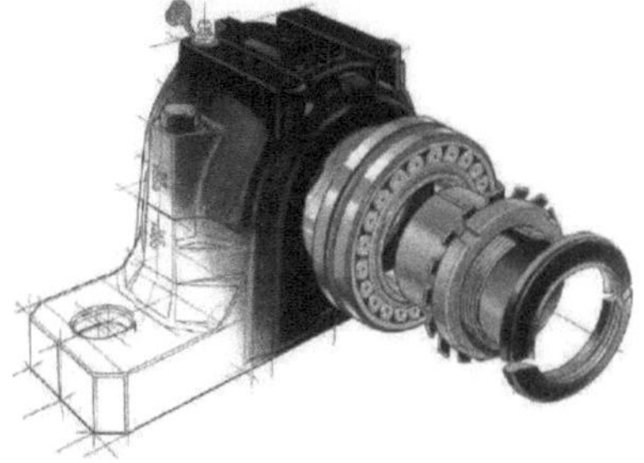

Figure 4.1 - Roller bearing with rotating shaft Source: Author.

According to (ANDRADE JÚNIOR, 1994) this type of bearing is used when greater speed and less friction are required. The behaviour of the rolling bearing can be checked by touch and hearing. However, in order to assess the rotation process, it is necessary to rotate the bearing slowly by touch, which allows you to see whether the movement is having difficulty rotating or not. When assessing by hearing, it is necessary to rotate the bearing at low speeds, taking into account the noise emitted, which can be classified as: scraping, rattling or metallic; if this occurs, the raceways are dirty, peeling, loose or lacking lubrication. The rolling bearing is made up of two concentric rings and between these rings are placed rolling elements such as: Ball, Roller and Needle. Figure 4.2 below shows the three types of bearing: ball, roller and needle:

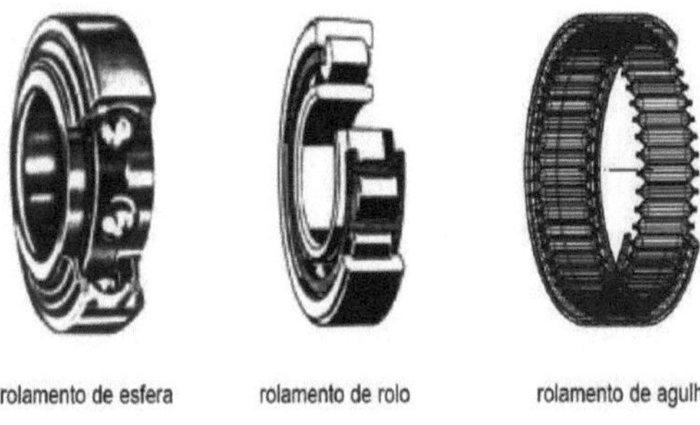

Figure 4.2 - Types of rolling bearings (Adapted from Andrade Junior, 1994).

4.2.1. Ball bearings

Ball bearings are suitable for higher speeds. The transmission of the ring with the ball is precise and therefore lubrication is essential, as the balls move in a single path between the rings, allowing free movement of the shaft. The main parts of a ball bearing arrangement according to General Motors Corp. are as shown in figure 4.3 below:

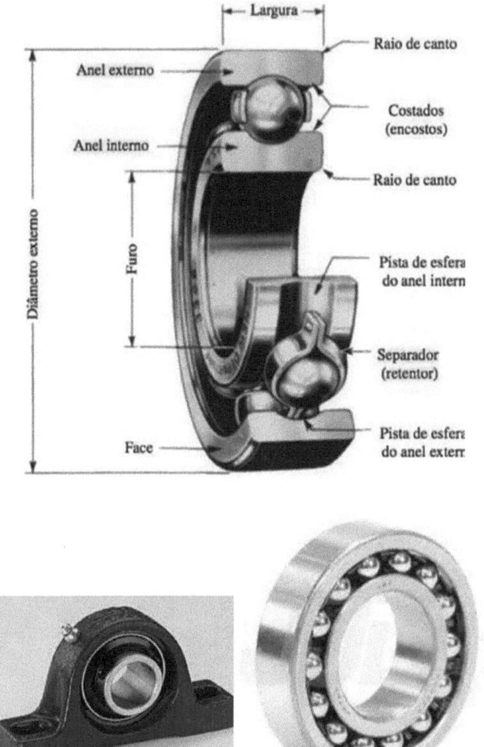

Figure 4.3 - Nomenclature of a ball bearing (Adapted from General Motors Corp).

4.2.2 Roller bearing

These are machine elements in the shape of conical or cylindrical rollers or barrels spread throughout the centre between the two rings like the balls, but they only allow the rollers to be fixed and are used when lower speeds and higher loads are required. (Silva, Santos and Fernandes, no date, p.5)

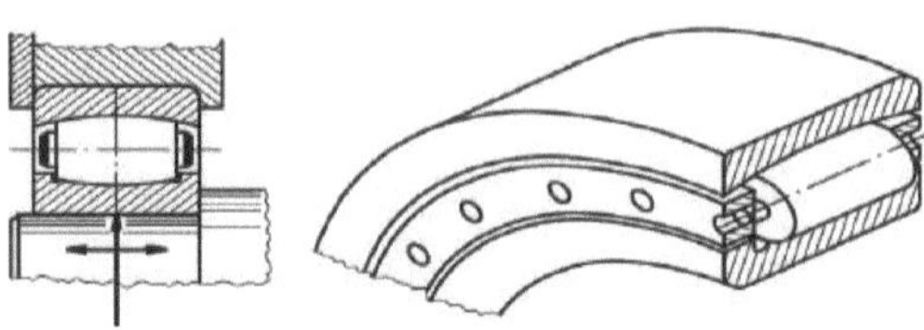

Figure 4.4 - Roller bearing (Adapted from Silva, Santos and Fernandes, no date, p.5).

4.2.3 Needle roller bearing

The rolling elements between the rings have small diameters of no more than 5mm and a length 3 to 10 times the diameter. These are used for oscillating mechanisms because needle bearings have high rigidity, where radial space is limited and the load is not constant.

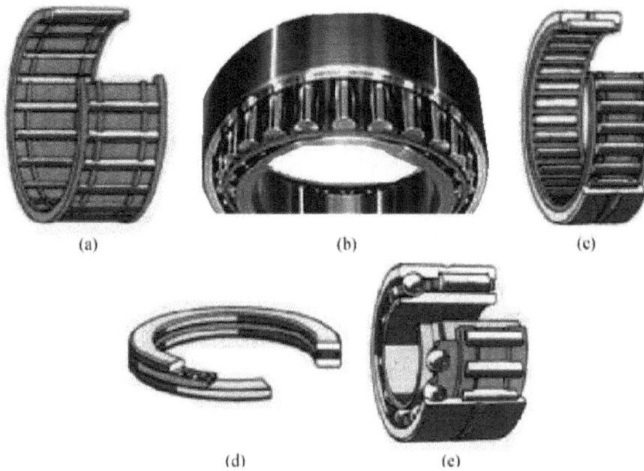

Figure 4.5 - Needle roller bearings: (a) single-row needle roller bearing; (b) single-row needle roller bearing; (c) bottomless needle roller bearing; (d) needle roller thrust bearing; (e) combined needle roller bearing (lecture notes).

The process of choosing a rolling bearing follows the flowchart in the figure below drawn up by (NSK bearings, 2014).

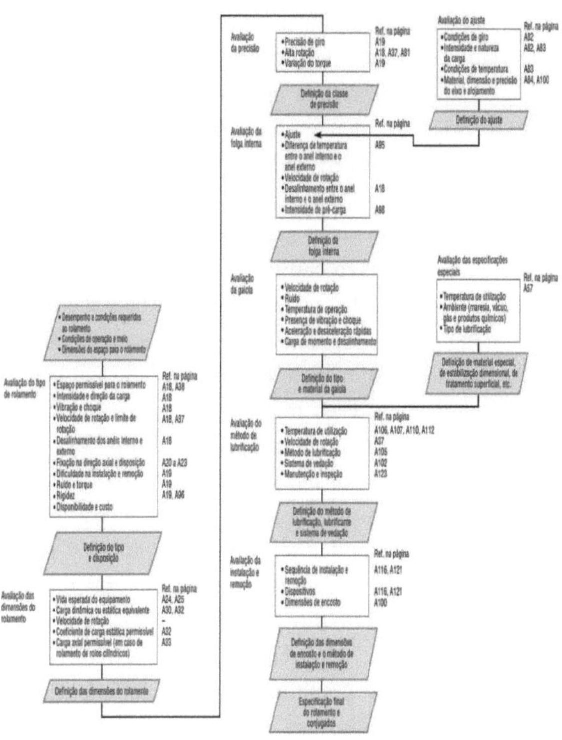

Figure 4.6 - Selection process for a rolling bearing (Adapted from NSK bearings, 2014).

4.2.4 Bearing material considerations

Cast iron was discovered in the mid 500s (AD) and was first commercialised in 1388. It has been the most common metal alloy used in tribological applications. Cast irons are alloys with wide application in mechanical engineering and the automotive industry, being used, for example, in the manufacture of internal combustion engine blocks or engine parts in general. In 2005, of the approximately 60% of ferrous metals produced in Germany (approximately 4 million tonnes), 2.5 million tonnes corresponded to the production of grey cast iron, followed by nodular cast iron, with almost 35% or 1.4 million tonnes. These alloys offer relatively low production costs, satisfactory mechanical properties and high resistance to corrosion at high temperatures (FRANCKLIN, 2009). Cast iron can be defined as an Fe-C alloy containing approximately 2%

carbon, often resulting in free carbon in the form of graphite particles. In this way, it is a material composed of graphite particles dispersed in a metallic matrix (Callister, 2003).Cast irons are classified into six classes: 1) grey cast iron; 2) white cast iron; 3) blended cast iron; 4) malleable cast iron; 5) compacted graphite cast iron and 6) nodular cast iron. Traditionally, cast irons are classified according to the colour of their fracture, as grey, white or mixed. A microstructural analysis shows that grey cast irons have graphite (C) in their composition, white cast irons have carbides (Fe3C, M3C or M7C3) and mixed cast irons a mixture of the two phases. Graphite can take the form of compacts, veins or nodules, among others, depending on the presence of small quantities of elements, the most important of which are magnesium and cerium added in a process known as nodulisation. Cast iron has particles of different shapes that directly affect its thermo-mechanical properties. Hardness and ductility are strongly dependent on the shape of the graphite particle. Particles with nodular shapes increase these properties, while more elongated particles or those with irregular contours are detrimental due to the concentration of stress points. Cast iron can therefore be classified according to the shape of its graphite particles (Chiaverini, 1990).Graphite formation in cast irons is regulated by the composition and cooling rate. Concentrations of silicon (Si) greater than approximately 1% promote the graphitisation (graphite formation). Very low cooling rates during solidification also favour the formation of graphite. The most common types of cast iron are grey, nodular (or ductile), white and malleable cast iron (MARMONTEL, 2011).

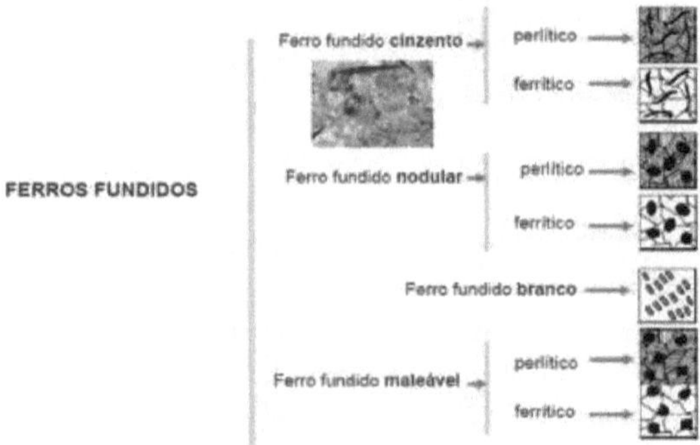

Figure 4.7 - Characterisation of different cast irons (Adapted from CABRAL, 2015).

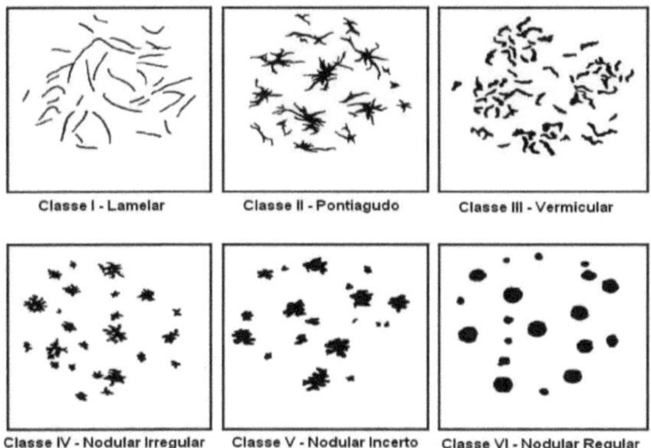

Figure 4.8 - Reference image for the six classes of graphite particles according to the ISSO-945 standard source: (Francklin, 2009).

In the work carried out by (Francklin, 2009), some structures similar to the one shown in figure 4.7 were observed, as shown in figure 4.8 where the microscopic image of a sample of the material in its raw state of fusion, where graphite nodules can be seen (A), and the same sample after chemical attack with Nital (composed of 2% nitric acid and 98% alcohol) (B), where graphite nodules can be seen in a metallic matrix.

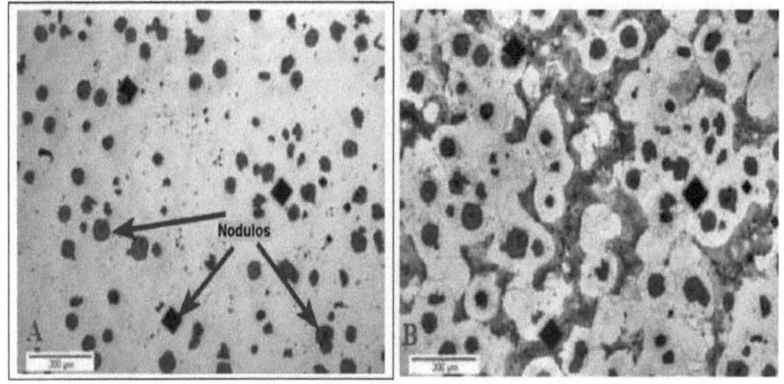

Figure 4.9 - Microscope of nodular cast iron sample Source: (Francklin, 2009).

As can be seen in Figures 4.10 (a), (b), (c) and (d), the morphology seen in the work by (Cabral, 2015), presented the typical microconstituents of a grey cast iron, i.e. pearlite, ferrite and graphite in a homogeneous form. The typical microstructure of grey cast iron is a pearlitic matrix with dispersed graphite veins. It can therefore be seen that the images in Figure 4.10 can be classified according to the type and distribution of graphite veins, in accordance with ASTM standards.

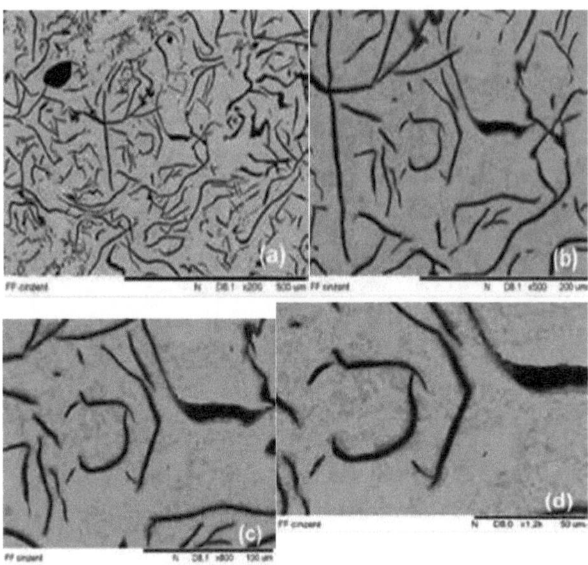

Figure 4.10 - Microscopy of grey cast iron: a) 200x; b) 500x; c) 800x and d) 1200x Source: (CABRAL et al., 2015).

Nodular cast irons have a microstructure, in the raw state of solidification, with graphite predominantly in the nodular or spheroidal form, as shown in Figure 4.11. In order for this modularity to be formed in the microstructure of the material, it is necessary for the cast iron to react with a modularising alloy. (FERREIRA et al., 2017)

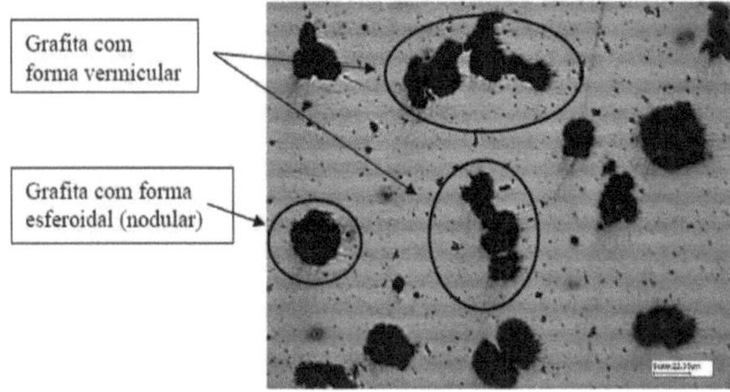

Figure 4.11 - Microstructure of a FE- 40015 cast iron. 2% nital attack (FERREIRA et al., 2017).

5. MACHINES ROTARY

According to (Ribeiro, 2007) all rotating equipment has certain levels of noise and vibration when in operation. A proportion of these vibrations are caused by mechanical defects or disturbing secondary excitations, which directly interfere with equipment performance. Whatever the increase in vibration level, it will be an indication of a worsening defect, be it an unbalance, a failing bearing, mechanical clearance above that specified by the design, etc.Ribeiro (2007) also mentions that careful monitoring of the vibrations of rotating equipment can indicate the main causes of deviations in dynamic behaviour, forming the basis of predictive maintenance techniques. (JR; SB, 2013) cites (GÓZ; SILVA, 2002). One of the main problems related to this type of machine is unbalance, since a rotating rotor generates dynamic forces that propagate to the parts of the machine that support it. The forces generated by unbalance, even if they are small, increase the workload of the machine's parts and, at the very least, reduce its useful life, thus causing losses for the entrepreneur. In addition to the damage caused, another consequence of unbalance is loss of quality, increased scrap, vibration and noise.

5.1. BEARING FAULTS

According to (Trevisan; Reguly, 2010), analysing the cause of bearing failures can be made much easier by using guides developed by the bearing manufacturers themselves. In these guides, you can find figures that illustrate the failure that has occurred, as well as the probable causes and ways to avoid damage. Figure 5.1 illustrates failures due to deviations in lubrication and/or lubricant. Both are the result of lengthy studies and are usually presented in a rapid diagnosis guide for bearing occurrences (NSK, 2001; SKF, 2005).

Figure 5.1 - Angular contact bearing inner ring (adapted from NSK, 2001).

Figure 5.2 shows the following symptom: Scaling around the circumference of the raceway. Cause: poor lubrication caused by cutting fluid entering the bearing.

Figure 5.2 - Spherical roller bearing inner ring (Adapted from NSK, 2001).

Figure 5.3a shows the symptom: exhaust in only one line. Cause: poor lubrication. In figure 5.3b, we have Symptom: peeling along the track. Cause: poor lubrication.

Figure 5.3 - Spherical roller bearing and inner ring of angular contact bearing (Adapted from NSK, 2001).

Figure 5.4 - Spherical roller bearing outer ring (Adapted from NSK, 2001).
Symptom: peeling has occurred near the edge. Cause: poor lubrication.

5.2 TECHNIQUES FOR IDENTIFYING FAULTS IN ROLLING BEARINGS

5.2.1 Ultrasound technique

In the article Analysing bearing wear using ultrasound-assisted lubrication published by trevisan and reguly in 2010, a detailed study is made of bearing failures that start at high frequencies, above the audible limit. In this way, by measuring the energy generated by the bearing in the ultrasonic region, it is possible to solve problems (especially those related to lubrication) well in advance, compared to the other predictive techniques used so far. According to (Trevisan; Reguly, 2010), this technique is mainly used in grease-lubricated engine bearings. It can be used to identify bearings with little or no lubrication and with advanced wear. It is also possible to prevent over-lubrication, which is the cause of many failures in grease-lubricated equipment, as it generates overheating, as well as external and internal leaks.

5.2.2 Vibration analysis technique

According to (Sanches 2005) vibration analysis is a powerful tool for diagnosing problems in machinery. There are many ways of obtaining vibration data and presenting it in order to detect and identify specific problems in rotating machinery. Sanches (2005) also says that, according to (GEITNER and BLOCH, 1994), the procedure of obtaining and presenting the vibration amplitudes for all the frequencies present is perhaps the most useful of all the analysis techniques. It is estimated that 85% of the problems that occur in rotating machinery can be identified using frequency domain analysis.(Sanches 2005) Instability in rotational dynamics is almost exclusively due to non-synchronous excitations. In fact, this is a characteristic of instability in turbo machines, due to the excitation of the rotor-bearing system at frequencies different from the rotor's rotational speed. The causes of instability in rotational dynamics are not associated with unbalance in the rotor, but are generally associated with non-linearity in the rolling bearing.The article entitled Bearing Defect Analysis Techniques - Traditional Technique, New Technology and Prospects for Use at Açominas, published by Vasconcelos and Ribeira in 2010, shows some of the traditional techniques inherent in lubrication and discusses how the evolution of analysis techniques has allowed defects to be detected at early stages throughout the speed range, helping maintenance to plan repairs and reducing costs due to any emergency equipment downtime. The authors state that bearing defects can be predicted by measuring vibration and monitoring the presence of defect frequencies and their multiples, so diagnosis should not only take amplitude into account. The fault frequencies are calculated taking into account the constructive characteristics of the bearings. The expressions used for the calculation are shown in figure 5.5 adapted from the article by (Vasconcelos and Ribeira, 2010).

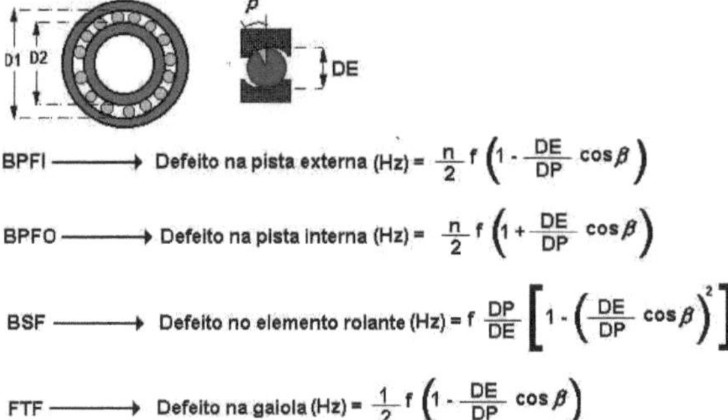

Figure 5.5 - Formulas for calculating bearing fault frequency (Adapted from Vasconcelos and Ribeira , 2010).

Where:

$$DP = \frac{D1+D2}{2}$$

(5.1)

Being:

n=number of rolling elements; f=frequency of rotation;

DE= rolling element diameter; DP= primitive diameter;

β= contact angle.

According to (Vasconcelos and Ribeira, 2010) a pattern of defects can appear, which is divided into 4 (four) phases, in phase 1, the first signs of a defective bearing will appear in the spectrum at high frequencies in multiples of the fundamental frequencies of the defect. The reason these high frequencies appear first is due to the excitation of the natural frequencies of the bearings or structure. The fundamental frequency (1 X BPFI) does not appear in this phase, as there is a specific frequency for each material and its harmonics. Figure 5.6 shows the type spectrum of this phase.

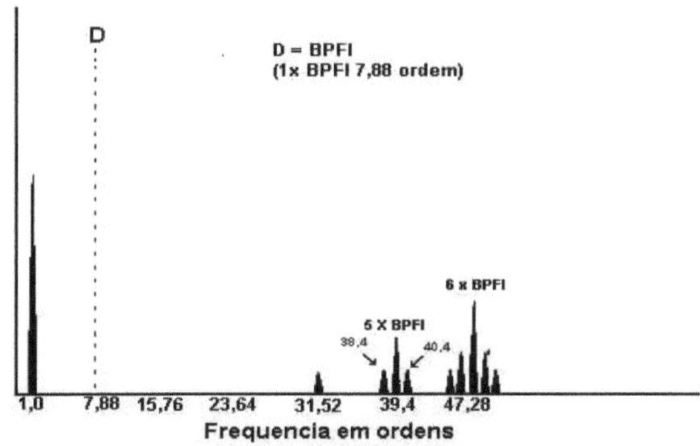

Figure 5.6 - Vibration spectrum - presence of multiple highs (Adapted from Vasconcelos and Ribeira, 2010).

In phase 2 (Vasconcelos and Ribeira, 2010) they explain that more harmonics of the fault frequency will appear in the spectrum. As degradation continues, shaft rotation will often appear, modulating the fault frequencies. The amplitude of the sidebands relative to multiples of the fault frequency is important at this point. Therefore, the amplitude of the sidebands exceeding the amplitude of the multiples of the fault frequency can indicate significant damage. As the bearing degrades, the fault frequencies that will appear in the spectrum will not be exactly the same as those calculated. This is because the degradation of the bearing will cause the internal geometry of the bearing to change. Figure 5.7 shows the spectrum of this phase.

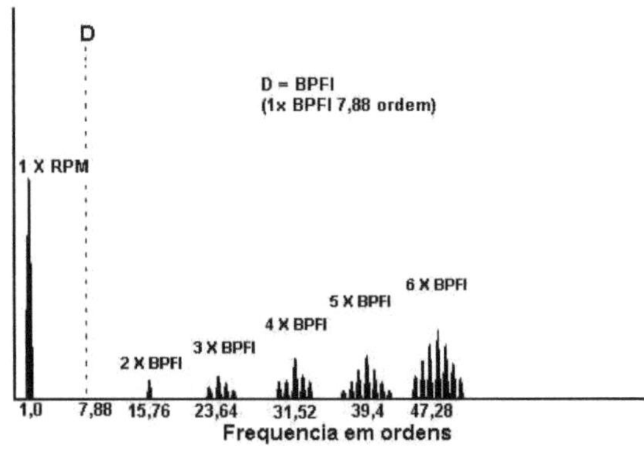

Figure 5.7 - Vibration spectrum - progression of the defect - appearance of multiple lateral bands of rotation around the harmonics of the BPFI (Adapted from Vasconcelos and Ribeira, 2010).

Phase 3 shows the presence of the fundamental fault frequency and its multiples, as well as sidebands. In addition, sidebands of the fault frequency of the rolling elements and cage can appear around the fault frequencies of the inner or outer race. This situation is usually found after advanced bearing degradation. The expected life of the bearing will depend on the rotation of the shaft and the bearing load. Figure 5.8 (Vasconcelos and Ribeira, 2010)

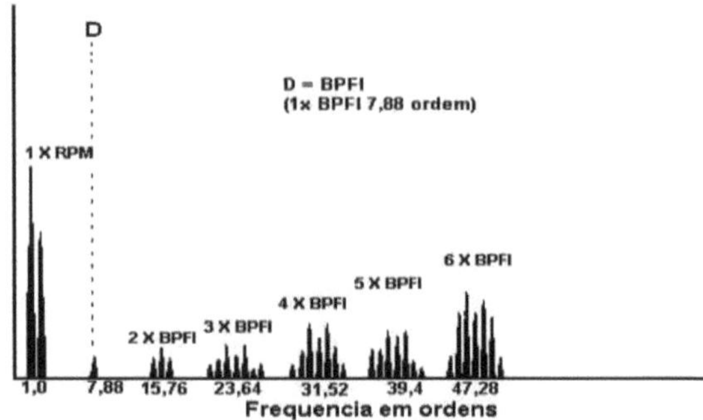

Figure 5.8 - BPFI fundamental frequency detected in the vibration spectrum (Adapted from Vasconcelos and Ribeira, 2010).

In phase four, as the bearing continues to degrade the internal clearances may increase, further accelerating the deterioration of the bearing components. This will allow more impacts to occur inside the bearing. As the impacts increase, they will appear in the spectrum as a rise in the tread. The peaks can be seen decreasing in amplitude and becoming less distinct as the tread increases. Failure is imminent at this point. It is important to realise that each bearing can develop different failure modes. Figure 5.9 (Vasconcelos and Ribeira, 2010).

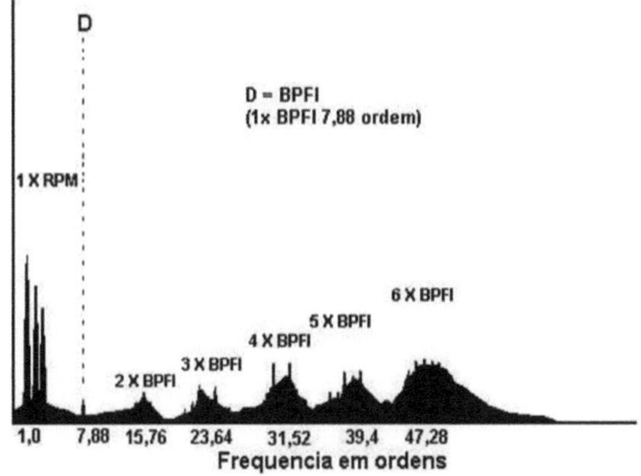

Figure 5.9 - Vibration spectrum altered by noise (Adapted from Vasconcelos and Ribeira, 2010).

5.2.3 Bearing fault detection technique - DEMODULATION

The Demodulation or Envelope technique as (Scheffer and Girdhar, 2004) points out: comprises as cited by (Nichterwitz, 2013), certain procedures that are applied to the signal. These are:

• Application of the Fourier transform to the signal (FFT), in order to identify a frequency range where there was an increase in relation to peak frequency measurements collected earlier. The elevation occurs due to the excitation of natural frequencies characteristic of the bearing or structure, due to the fault in the bearing.

• Application of a band-pass or low-pass filter to the signal in order to eliminate high-amplitude low frequencies, which are generally related to misalignment and unbalance.

• Application of the Hilbert transform in order to obtain the envelope, figure 5.11, of the fault signal.

• Application of the FFT to the envelope in order to obtain the frequencies of the defects.

All these procedures can be illustrated in figure 5.10 below.

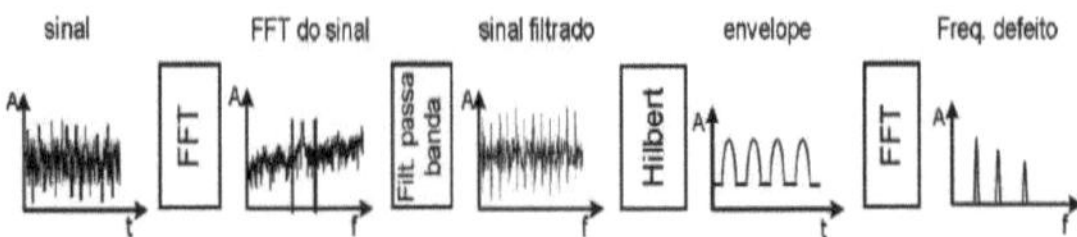

Figure 5.10 - Procedure adopted in the demodulation methodology (Adapted from Bezerra and Pederiva, 2004).

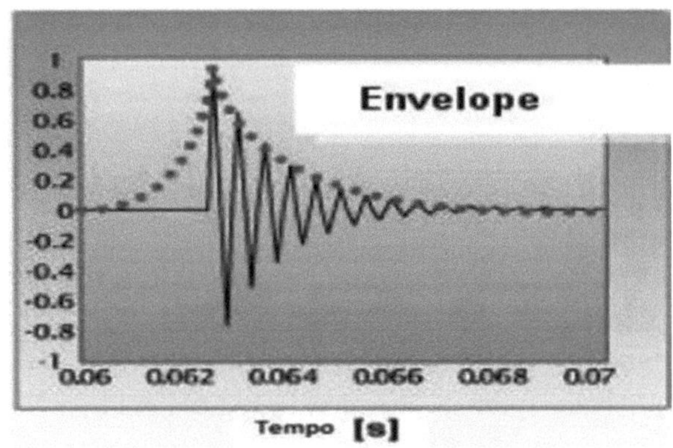

Figure 5.11 - Peak value within a sample captured at a Δt(Adapted from Saavedra and Estupiñan, 2002)

5.2.4 Bearing fault detection technique - PEAKVUE

As mentioned (Scheffer and Girdhar, 2004), the Peak Value Technique, like demodulation, isolates the resonance zones by passing through a high-pass or band-pass filter to remove low frequencies. The signal then passes through an analogue-to-digital converter, each sample is analysed and only amplitude levels that exceed a specified value are assigned digital signals. In other words, if the analyser is programmed to take 100 samples of the analogue signal and there are strong impacts, it will obtain 100 digital peak values, Figure 5.12. The digital waveform in the time domain will only have the highest positive values, one per sample. If there are no defects, there will be no high pulses, and only low-amplitude noise (from the signal or instrument) will exist and will not be transformed into a digital signal. A pulse from an impact due to a bearing or gear defect has a very short time duration and tends to be periodic, so there will only be high peak values in the samples where the pulse occurs (Nichterwitz, 2013).Finally, the FFT (Fourier transform) algorithm processes the digital information and builds the spectrum, which shows only the fundamental peaks and harmonics with frequencies equal to those of the pulses. As (Nichterwitz, 2013) puts it, with PeakVue, the signal is extracted from high-frequency data, greater than or equal to the Nyquist frequency, but it is not rectified or enveloped by a low-pass filter. The digital data block then consists of the maximum absolute values of the waveform in the time domain, at each time increment defined by the sampling rate. Therefore, the analysis of the representative values of the waveform in time is the analysis of the peak values.

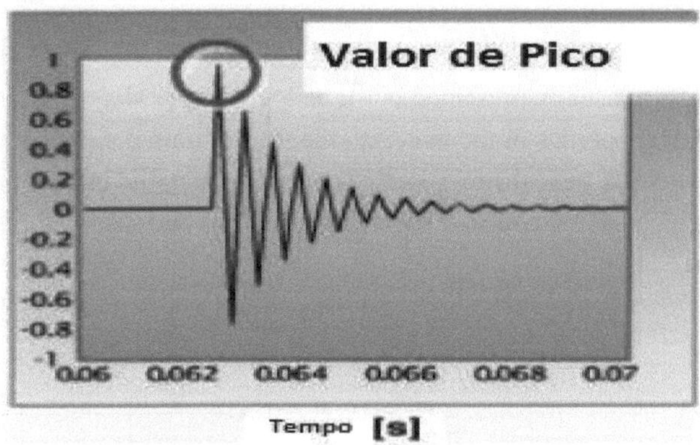

Figure 5.12 - Peak value within a sample captured at one Δt (Adapted from Saavedra and Estupiñan, 2002)

5.2.5 Mechanical characterisation technique

Tensile test

For (Garcia and Santos, 2012) the tensile test is based on direct measurements of stresses and strains, i.e. it consists of mechanically subjecting a sample through the application of tension (σ) at low load application speeds and the modulus of elasticity and shear is determined from the slope o f the linear region of the stress x strain diagram. Other mechanical properties such as yield strength, yield strength and elongation are also determined by the relationship of the stress x strain diagram. Equation 5.2 and 5.3 are called stress σ(t) and strain ε(t), both calculated as a function of time.

Where:

$$\sigma(t) = \frac{F(t)}{A} \tag{5.2}$$

$$\varepsilon(t)=\frac{\Delta L(t)}{L} \quad (5.3)$$

σ(t) - is the engineering voltage over time;

F(t) - is the force applied axially to the specimen; A - cross-sectional area of the specimen;

ε(t) - is the engineering deformation over time; ΔL - Elongation;

L - Initial length.

In the international system (SI), the unit used for stress is the Pascal (1 Pa = $1N/m^2$). Stress is usually presented in MPa (1 MPa = 10^6 Pa). In order to obtain the mechanical properties of the material, the stress-strain diagram is used. Figure 5.13 shows a typical stress-strain curve for a metallic material, tested uniaxially, which has two distinct regions, one elastic and the other plastic.

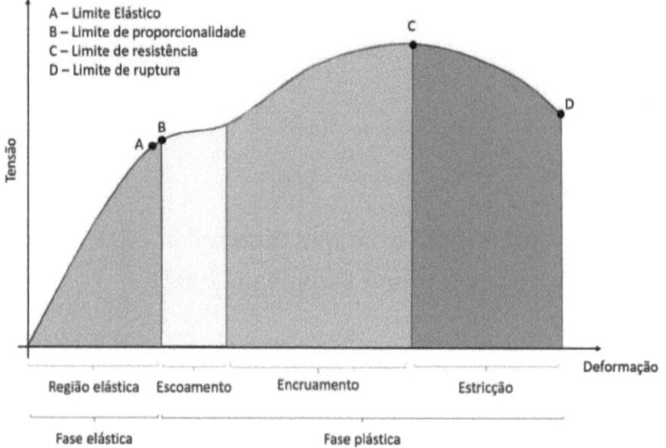

Figure 5.13- Typical diagram of the stress and strain relationship (Adapted from the biopid website, tensile test, no year).

Young's modulus, a mechanical parameter measured in the elastic region, measures the rigidity of the material. It is related to the elastic deformation zone of materials and is defined by Hooke's law, represented by equation 5.4:

$$\sigma(t)=E \times \varepsilon(t), \ \forall \sigma(t) < \sigma_e \quad (5.4)$$

Hooke's law is measured in the linear elastic regime where the deformation suffered by the solid is totally reversible when the load is removed. In this region, the elastic modulus can be obtained from the angular coefficient of the graph of stress as a function of strain, as shown in figure 5.14.

Figure 5.14 - Stress-strain curve in the elastic regime (Adapted from Paiva, 2001).

In the master's thesis entitled: A Influência da variação do módulo de elasticidade na previsão computacional do retorno elástico em aço de alta resistência. developed by Lajarin in 2012 at the Federal University of Paraná. Lajarin mentions that the elastic modulus parameter is represented by the slope tangent of the secant line formed by the connection between the maximum and minimum stress points of the unloading curve. The purpose of the tensile test is to find out the material's capacity to withstand stress when an effort is applied that tends to elongate it in a uniaxial direction, constantly over time until it breaks. This is how mechanical properties such as tensile strength, yield strength, modulus of elasticity, elongation, among others, are quantitatively determined. According to (Callister, 2003) as an axial stress (σ) is applied, a negative ratio arises between the lateral and axial stresses, which is defined as Poisson's ratio, μ, as shown in equation 5.5:

$$\mu = -\frac{\varepsilon_x}{\varepsilon_z} = -\frac{\varepsilon_y}{\varepsilon_z} \qquad (5.5)$$

5.2.6 Metallographic Technique

For (Junior, Miranda and Cunha, 2018) the procedure for analysing the effect of normalising and annealing heat treatment is based on its microstructure and is carried out through metallography in order to guarantee changes in its mechanical properties. According to (Colpaert, 2008): cutting the sample, sanding, chemical attack and image acquisition using a microscope are crucial stages for analysing the microstructure of the part. A Figure 5.15 shows the preparation developed by (Junior, Miranda and Cunha, 2018).

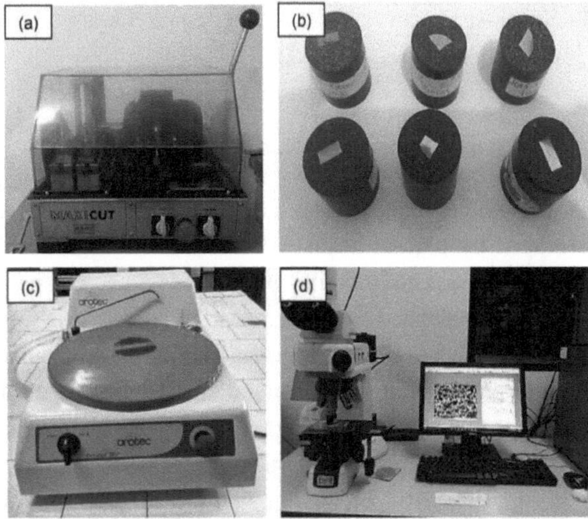

Figure 5.15 - Metallographic sample preparation procedure. (a) Polycutting machine used to remove the specimen; (b) Metallographic sample embedded; (c) Polishing machine used for grinding and polishing; (d) Acquisition of images in the optical microscope on samples after chemical attack (Adapted from Junior, Miranda and Cunha, 2018).

6. METHODOLOGY

The research will cover a study of techniques for identifying faults in rolling bearings. It will analyse and discuss ultrasound techniques, vibration analysis and metallurgical analysis based on the faults presented in rolling bearings. In the initial stage, an approach will be made to the literature with an exposition of other work carried out on this subject, with the research being done in books, articles and other TCCs. As we do not have the equipment for experimental analysis, an action plan will be drawn up for carrying out the tests, describing the tests by technique and the equipment normally used in these cases.

6.1 MATERIAL

The material used as a rolling bearing is usually cast iron, and the tables below show some of its mechanical and chemical properties:

Table 6.1 - Mechanical properties of grey cast irons (Adapted from Marcondes, no ano, p.8).

ASTM A 48 class	Resistência à tração		Resisistência à torpão		Resistência à compressão		Limite de fadiga em dobramento		dureza (HB)
	MPa	ksi	MPa	ksi	MPa	ksi	MPa	ksi	
20	152	22	179	26	572	83	69	10	156
25	179	26	220	32	660	97	79	11.5	174
30	214	31	276	40	752	109	97	14	210
35	252	36.5	334	48.5	855	124	110	16	212
40	293	42.5	393	57	965	140	128	18.5	235
50	362	52.5	503	73	1130	164	148	21.5	262
60	431	62.5	610	88.5	1293	187.5	169	24.5	302

Table 6.2 - Chemical properties of grey iron according to ASTM (Adapted from infomet, Classification of cast irons, no year).

Classe ASTM	Composição química (%)				
	C	Si	Mn	P	S
20	3,10/3,80	2,20/2,60	0,50/0,80	0,20/0,80	0,08/0,13
25	3,00/3,50	1,90/2,40	0,50/0,80	0,15/0,50	0,08/0,13
30	2,90/3,40	1,70/2,30	0,45/0,80	0,15/0,30	0,08/0,12
35	2,80/3,30	1,60/2,20	0,45/0,70	0,10/0,30	0,06/0,12
40	2,75/3,20	1,50/2,20	0,45/0,70	0,07/0,25	0,05/0,12
50	2,55/3,10	1,40/2,10	0,50/0,80	0,07/0,20	0,06/0,12
60	2,50/3,00	1,20/2,20	0,50/1,00	0,05/0,20	0,05/0,12

6.2. EQUIPMENT

In order to use the ultrasound fault identification technique in rolling bearings, as mentioned in (Trevisan; Reguly, 2010), an ultrasound data collector similar to the device shown below was used to take the measurements.

Figure 6.1 - Illustration of the SDT 170MD device (Adapted from the SDT 170 website, 2010).

Figure 6.2 - Illustration of the accessories used to monitor the condition of rolling bearings. From left to right: (i) pen-type sensor; (ii) junction for threading the accelerometer and coupling the grease pump; (iii) accelerometer; (iv) magnetic sensor; (v) extension cable; (vi) collection device: (vii) headphones (Adapted from Braskem/Triunfo, 2010).

Also according to (Trevisan; Reguly, 2010) To read the sound, the piezoelectric crystal picks up the ultrasonic waves received, and as it has the thickness calculated to resonate at the aforementioned frequency, it then changes its electrical properties and generates a proportional voltage signal. The data collected is then presented using the voltage signal generated by the ultrasonic sensor, but in decibels (dBμV). This is a dimensionless measure of the ratio between two voltages (μV) and is obtained from equation 1 (Murphy and Rienstra, 2009). The definition of dB is obtained using the logarithm, which makes it possible to replace multiplication and division with addition and subtraction. Table 6.3 shows a brief conversion between the dBμV measurement and the voltage μV.

$$x_{db} = 20 \log_{10} X \quad (6.1)$$

Being:

X_{db} = Measure of the ratio between X and X_O ;

X = Measured voltage value;

X_0 = Reference voltage value (1 μV).

Table 6.3 - Conversion of dBμV values (Adapted from the SDT 170 website, 2010).

40 dBμV	20 dBμV	6 dBμV	0 dBμV	-6 dBμV	-20 dBμV	-40 dBμV
100 dBμV	10 dBμV	2 dBμV	1 dBμV	0.5 dBμV	0.1 dBμV	0.01 dBμV

(Trevisan; Reguly) states that: In order to interpret the data in Table 6.3, it must be verified that the value of 6 dBμV , for example, is equivalent to doubling the voltage value μV , as it changes from 1 to 2 μV . This same concept must be applied to all the values shown. Similarly, an increase of 40 dBμV is equivalent to a 100-fold increase in the voltage generated. In order to use the technique of vibration analysis, which involves capturing signals, according to Sequeira (no date, p.4) "[...] sensors are called mechanical vibration transducers. There are various types of sensor, with the accelerometer being the most widely used due to its enormous versatility, while other sensors are restricted to very specific applications." In addition, the device used to analyse vibrations is called a vibration analyser. This same piece of equipment is popularly referred to as a "vibration collector" among maintenance staff and technicians who use it (Junior, 2018). Figure 6.3 shows an example of a portable signal analyser.

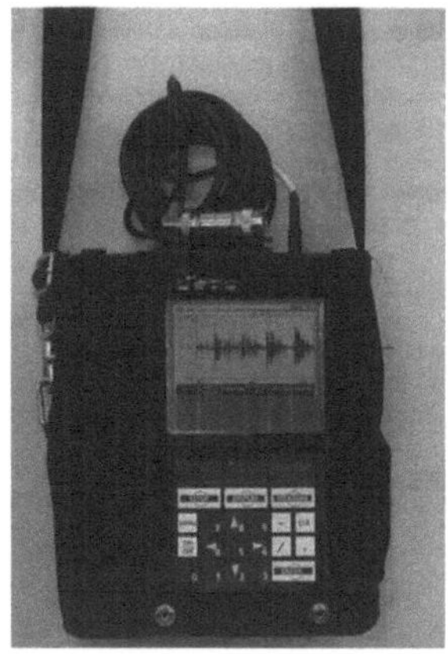

Figure 6.3 - Portable signal analyser (Adapted from WEBER et al. , 2009).

(Junior, 2018) in his final course work, entitled "Vibration Analysis in Industrial Bearings", carried out a case study using vibration analysis on local industrial equipment to detect faults in axial, vertical and horizontal motor bearings. (Junior, 2018) used the following equipment:

a) PRUFTECHNIK VIBXPERT II vibration analyser serial no. 32017;

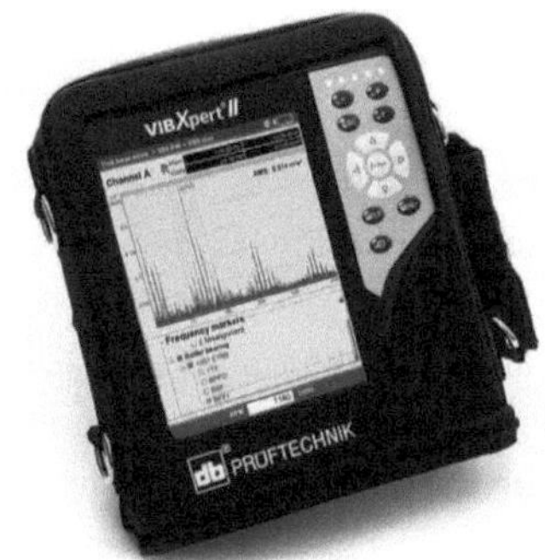

Figure 6.4 - PRUFTECHNIK VIBXPERT II vibration analyser (Adapted from PRUFTECHNIK, 2017)

b) Acele mV/g

80g dynamic, working temperature from -40 to 130ºC, application range from 0.6Hz to 15kHz with various interface configurations, 2-pin military connector, integrated cable made of silicone, polyurethane, Teflon-braided stainless steel and Teflon with stainless steel armour.

Figure 6.5 - RPM-SR110 accelerometer (Adapted from RPM SUL, 2018).

c) Use of OMNITREND-PRUFTECHNIK software to analyse the data collected.

(Junior, 2018) In equipment with several bearings, the measurement sequence is determined from the driving equipment to the driven equipment, i.e. following the direction of force (from the rear bearing of the motor to the rear of the driven equipment), as shown in Figure 6.6.

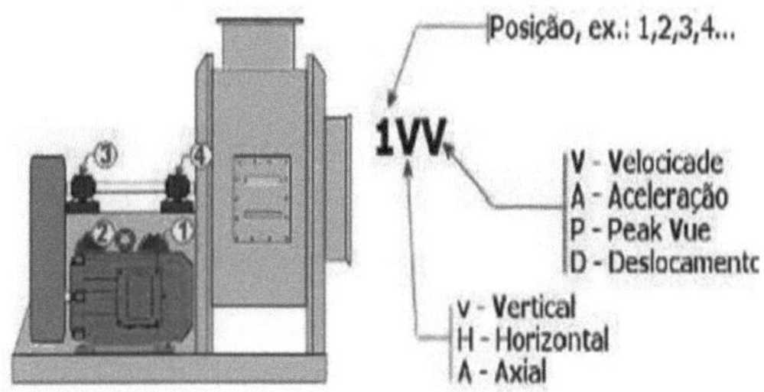

Figure 6.6 - Measurement procedure (Adapted from RPM SUL, 2016).

(Junior, 2018) also mentions that the following abbreviations were used:

a) LA = Driven Side (used for rollers) or Coupled Side (used for motors, gearboxes, pumps, etc.);

b) LOA = Side Opposite Drive or Side Opposite Coupling;

c) LC = Command side;

d) LP = Pulley side;

e) LOP = Lateral Opposite Pulley.

As far as classification is concerned, the colour criteria used to carry out the analyses are divided into four classes according to the levels of vibration and the characteristics found:

a) good: operating status is good (no faults, low vibration levels);

b) acceptable: the operating state is acceptable for a long period of operation;

c) attention: the operating state is not acceptable for a long period of operation;

d) critical: vibration levels can cause unexpected breakage;

e) unmonitored: unmonitored equipment.

Table 6.4 below, as pointed out by (Junior, 2018), shows the state of the machine taking into account the ISO 10816-3:2009 regulatory standard.

Table 6.4 - Criteria for judging the state of machinery (Adapted from RPM SUL, 2016).

Nível de Vibração (mm/s)	CRITÉRIO PARA JULGAMENTO DO ESTADO DE MÁQUINAS			
	até 20CV	de 20CV até 100CV	>100CV Base Rígida	>100CV Base Flexível
0,28	Bom	Bom	Bom	Bom
0,45	Bom	Bom	Bom	Bom
0,71	Bom	Bom	Bom	Bom
0,12	Aceitável	Bom	Bom	Bom
1,8	Aceitável	Aceitável	Bom	Bom
2,8	Atenção	Aceitável	Aceitável	Bom
4,5	Atenção	Atenção	Aceitável	Aceitável
7,1	Crítico	Atenção	Atenção	Aceitável
11,2	Crítico	Crítico	Atenção	Atenção
18	Crítico	Crítico	Crítico	Atenção
28	Crítico	Crítico	Crítico	Crítico
45	Crítico	Crítico	Crítico	Crítico

The equipment analysed by (Junior, 2018) was a motor and gearbox drive unit. The engine has the following characteristics:

a) model: MGI 500B;

b) RPM: 892 rpm to 1830 rpm

c) LOA bearing: 6322

d) LA bearing: NU 322

e) power: 951kW.

The gearbox has the following characteristics:

a) input shaft pinion: 21 teeth;

b) intermediate shaft crown: 79 teeth;

c) intermediate shaft pinion: 25 teeth;

d) output shaft crown: 82 teeth;

e) input shaft bearing: 23138 CC W33;

f) intermediate shaft bearing: 23144 CC W33;

g) output shaft bearing, input side: 23068 CC W33;

h) output shaft bearing, spindle side: 23084 CC W33;

Rotations and frequencies were then calculated, giving rise to the following table
$Z1 \times N1 = Z2 \times N2$ = Gearing frequency 121 teeth x 18.10 Hz = 380.10 Hz
$N2$ = 380.10Hz / 79 teeth = 4.81 Hz
$Z3 \times N2 = Z4 \times N3$ = Gearing frequency 225 teeth x 4.81 Hz = 120.28 Hz
$N3$ = 120.28Hz / 82 teeth = 1.467 Hz

Table 6.5 - Calculation of rotations and frequencies (Adapted from Junior, 2010).

(RPM) Input	1086
(Hz) Input	18,10
(RPM) Intermediate	288,68
(Hz) Intermediate	4,81
(RPM) Output	88,01
(Hz) Output	1,467
Frequency (Hz) Gear 1	380,10
Frequency (Hz) Gear 2	120,28

As for parameterisation, vibration samples were taken at the points as described below:

a) 1 - Vertical LOA Motor Bearing (rear)

b) 2 - Horizontal LOA Motor Bearing (rear)

c) 3 - Horizontal Motor Bearing LA (front)

d) 4 - Axial Motor Bearing LA (front)

e) 5 - Vertical Bearing LA Input Reducer

f) 6 - Horizontal Bearing LA Input Reducer

g) 7 - Axial Bearing LA Reducer Input Shaft

h) 8 - Vertical Bearing Motor Side Output Shaft Reducer

i) 9 - Horizontal Bearing Motor Side Shaft Output Reducer

j) 10 - Axial Bearing Motor Side Output Shaft Reducer

k) 11 - Horizontal Bearing Side Spindle Shaft Output Reducer

The figures below show the equipment used in the study.

Figure 6.7 - Vibration collection points on the gearbox (Adapted from Junior, 2018).

Figure 6.8 - Vibration collection point (Adapted from Junior, 2018).

Figure 6.9 - Vibration collection point (Adapted from Junior, 2018).

According to (Junior, 2018), the bearing operated 24 hours a day, and according to the Chinese manufacturer DWR, it had a service life of 40,000 hours, or about 4 years and 7 months. The company's preventive plan was to replace the bearing every 4 years or around 35,000 hours. (Junior, 2018) Starting the analysis, according to the trend shown in Figure 6.10, there was an evolution in the acceleration and envelope levels of the gearbox input bearing, indicating bearing wear characteristics.

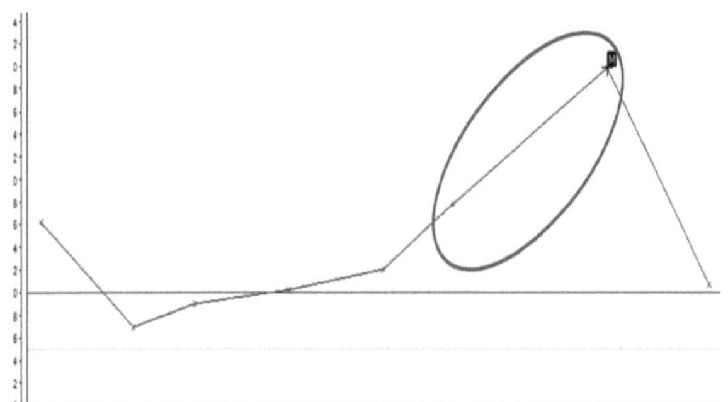

Figure 6.10 - Trend indicating bearing wear characteristics (Adapted from Junior, 2018).

(Junior, 2018) The signal in time, in orange, from point 3H (reducer inlet) shows the measurement before the evolution, still showing low amplitude and no evidence of peaks with repeating intervals.

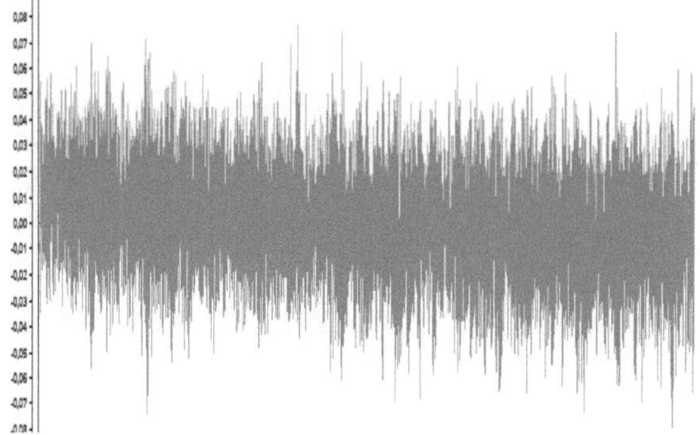

Figure 6.11 - Signal in time showing the measurement before the evolution (Adapted from Junior, 2018).

According to Junior (2018), the red time signal of point 3H shows defined peaks with the same impact interval, indicating the presence of a fault.

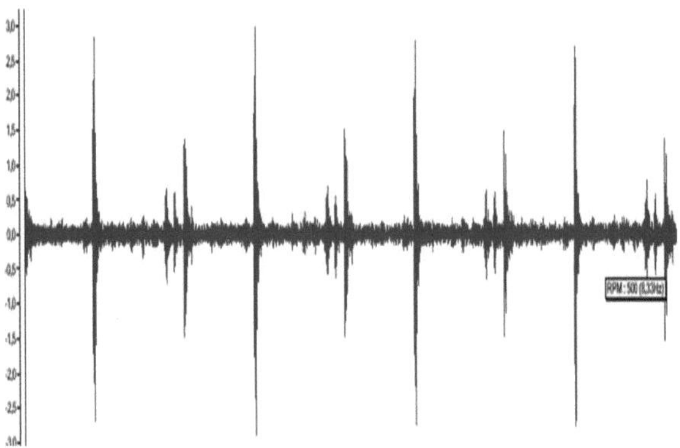

Figure 6.12 - Signal in time indicating the presence of a fault (Adapted from Junior, 2018).

Also in this same study, the acceleration spectrum of point 3H, in Figure 6.13, indicates lubricant film breakdown due to the "carpet" region in the 4000 to 7000 Hz range and indicates bearing wear due to the presence of defined peaks and a periodic interval in the 1000 to 2500 Hz range (Junior, 2018).

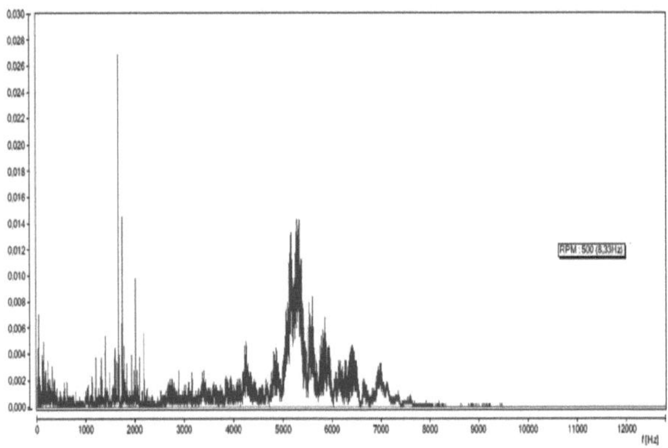

Figure 6.13 - Acceleration spectrum indicating lubricant film breakdown (Adapted from Junior, 2018).

(Junior, 2018) Figure 6.14 shows the acceleration spectra of point 3H in cascade, showing the evolution of the levels over time. The spectra shown are from the last collections made, with intervals of 1 month between each collection, i.e. a monthly collection cadence.

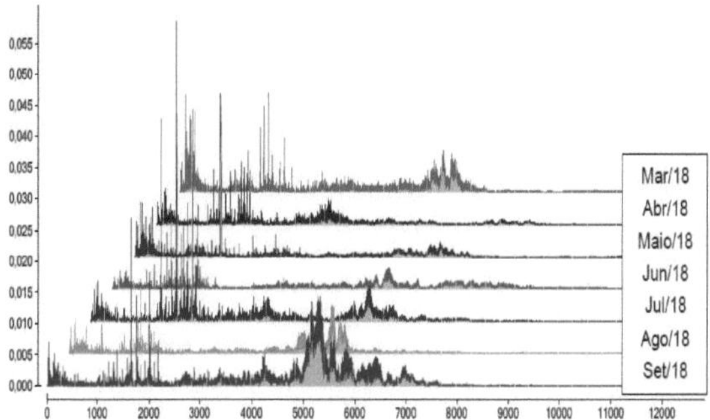

Figure 6.14 - Cascade acceleration spectra (Adapted from Junior, 2018).

The acceleration envelope spectrum of point 3H in Figure 6.15 indicates bearing wear due to the presence of defined peaks and a periodic interval (Junior, 2018).

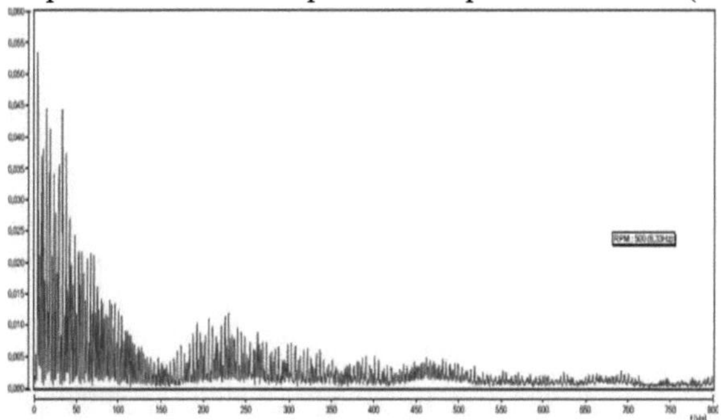

Figure 6.15 - Acceleration envelope spectrum indicating bearing wear (Adapted from Junior, 2018).

Figure 6.16 and Figure 6.17 show the progression of bearing wear and indicate that the bearings should be replaced.

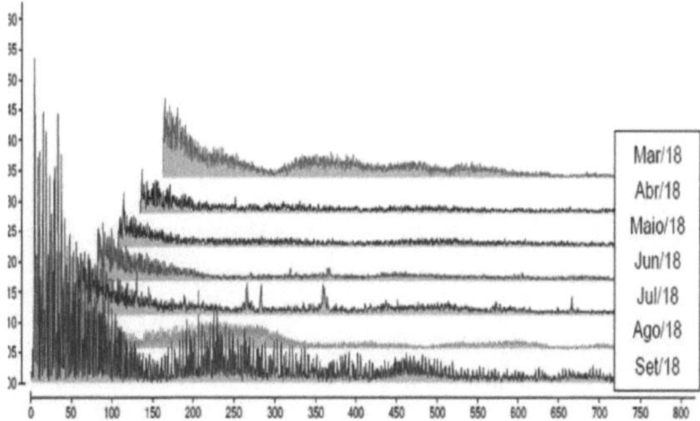

Figure 6.16 - Cascade acceleration envelope spectra (Adapted from Junior, 2018).

(Junior, 2018) As the failure frequencies of the bearing components are known, it is possible to register them in the OMNITREND-PRUFTECHNIK analysis software and superimpose these frequency markers on the acceleration envelope spectrum.

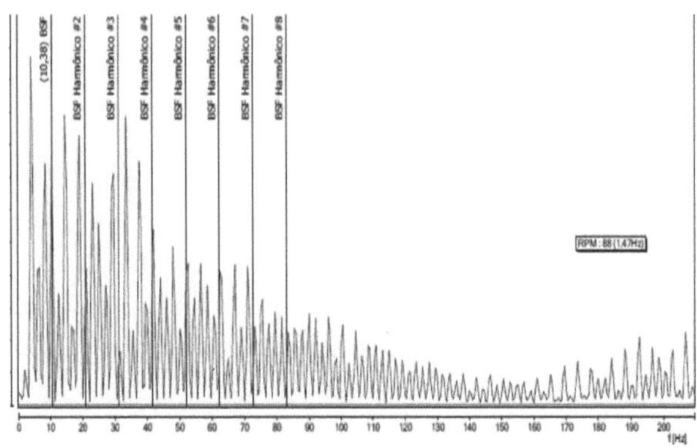

Figure 6.17 - Envelope spectrum of the 3H point (Adapted from Junior, 2018).

According to (Junior, 2018) In the envelope spectrum of point 3H in Figure 6.17, it is possible to see the existence of rotating element wear (BSF) and inner race wear (BPFI). Due to the amplitude, it can be said that there is already severe wear on one of the components.

(Junior, 2018) due to the high level of vibration and the evolution of the trend, added to the failure characteristics of the bearing components, with approximately 26,300 hours of operation, or about 3 years, it was recommended that it be replaced.

In the procedure for the metallographic technique, as already mentioned in sections 4.2.4 and 5.2.6, the following materials are generally used according to (Francklin, 2009) and (Junior, Miranda and Cunha, 2018):

a) Polycutting machine used for removing specimens;

b) Inlaid metallographic sample;

c) Polishing machine used for sanding and polishing;

d) Nital chemical attack (composed of 2 per cent nitric acid and 98 per cent alcohol);

e) Optical microscope for acquiring images of samples after chemical attack.

7. EXPECTED RESULTS

With this research we hope to acquire knowledge about the techniques used to predict possible causes of failure in equipment used for bearings on the shop floor, since it is a bibliographical study.

8. CONSIDERATIONS FINAL

Although this work is based only on a bibliographical study, with all the study being of the most common fault identification techniques used in rotating machinery in rolling bearings, even so, according to the theory addressed, the great importance and need for prior fault identification in rolling bearings is perceptible, since all the techniques, whether ultrasound, vibration, mechanical and metallographic characterisation, can considerably prevent damage to both the equipment and a production line. The work also made it possible to approach the principle of predictive maintenance, since this maintenance serves to predict and anticipate defects in the behaviour of machinery, providing better use and minimising defects that could reduce its useful life.

REFERENCES

GEITNER, Fred K.; BLOCH, Heinz. **Analysing and solving faults in mechanical systems**. Elsevier Brazil, 2016.

COLLINS,j.a.. **mechanical design of machine elements: a failure prevention perspective** Rio de Janeiro. LTC, 2015

NSK. 2001. NSK BEARING DOCTOR. **Rapid Diagnosis of Bearing Occurrences.**São Paulo, 36 p.

Rolling bearings. Available at <http:// www.essel.com.br/cursos/material/01. Accessed on: 25/02/2021.

TEXACO: **Ball and roller bearings**. 1964.

RIBEIRO MENNA, A, 2007. Master's thesis **"Fault detection in rolling bearings by broadband vibration analysis: a practical case of application in a population of rotating equipment"**, UFRS, Porto Alegre, Brazil.

JR, Berton. S.B,Roberto. **Problems related to rotating machinery**, 2013.

GARCIA, A.; SPIM, J. A.; SANTOS, C. A. dos. Materials testing. Rio de Janeiro: LTC, 2012.

TREVISAN, Renato; Reguly, Afonso. **Analysing wear in rolling bearings using ultrasound-assisted lubrication.** Estudos Tecnológicos - Vol. 6, n° 3: 122-139 (Sep/Dec 2010) doi: 10.4013/ete.2010.63.02

SACHES GARCIA, Mauricio, 2005. Master's thesis **"Analysing defects in rotating mechanical systems using vibration monitoring."** UFRJ, Rio de Janeiro, Brazil.

Ludwig, G.A. **Tests Performed by the builder on new products to prevent failure, loss prevention of rotating machinery.** new York: N.Y. 10017: the american society of mechanical engineers, 1972, p.3.

Yunus A. Çengel, John M. Cimbala. **Fluid Mechanics**. [S.l.]: MacGrawHill. 483 pages.

Robert L. Norton (2000). **Machine Design: An Approach**. [S.l.]: BookMan. 552 pages.

FRANCKLIN, Alexandre Reis **a brief study of nodular cast iron**. UEZO, Rio de Janeiro, 2009

CALLISTER, W.D. **Materials Science and Engineering: An Introduction.** 6th Edition. New York: 2003.

CHIAVERINI, V. **"Steels and cast irons"**. 6th ed., São Paulo: Brazilian Association of Metallurgy and Materials, 1990. 576 pgs.

CABRAL, Ricardo de Freitas et al. Evaluation of the microstructure and microhardness of grey cast iron. **Cadernos UniFOA**, Volta Redonda, n. 28, p. 17-23, Aug. 2015.

MARMONTEL, C. F. F; SILVA, J. M. G. G.; OLIVEIRA, L. L.; POLIONI, M. C. Analysis Metallography of Metals. Art and Science; 2011.

RAPHAEL Ferreira, EDUARDO Dalmolin, DIEGO Rodolfo S.L.. **Analysis of the microstructure and mechanical properties of fe-40015 nodular cast iron as a function of casting time** SECITY 2017 Santa Catarina

COLPAERT, H. **Metallography macrography and micrography of common steel products**, Bulletin 40, IPT, São Paulo, 2008.

JUNIOR, A.S.N; MIRANDA, Rodrigo da S.;CUNHA, A.P.A**. Analysis of mechanical and microstructural properties of SAE 1020 steel** Mechanical and industrial engineering Ponta grossa (PR) Atena Editora, 2018

VASCONCELOS, Nivaldo; RIBEIRO, Marcos Pinto **bearing defect analysis, traditional technique, new technology and prospects for use at Açominas.** Maintenance Engineer. AÇOMINAS

ANDRADE JÚNIOR, Irajá Gaspar. **Rolling Technology.** São Paulo: Brazilian Pulp and Paper Technical Association, 1994 32p.

LAJARIN, S. F. **Influence of the variation of the modulus of elasticity on the computational prediction of the elastic return in high-strength steels.** Federal University of Paraná 2012.

BEZERRA, R. A., Pederiva, R.; **"Fault Detection in Bearings by Vibration Analysis"**, Doctoral Thesis, State University of Campinas, Campinas - S.P. Brazil, 2004.

SAVEEDRA, P., Estupiñan, E. **"Diagnostic Techniques for the Vibration Analysis of Bearings"**, Internal document, Universidad de Tarapacá y Universidad de la Concepción, Chile, 2002.

NICHTERWITZ, M.P.; **"Comparative Study Between Peakvue and Envelope Vibration Signal Demodulation Methods for Bearing Failure Prediction"**, UFRS, Rio Grande do Sul - RS. Brazil, 2013.

Silva, Débora A.C; Santos, Érika B; Fernandes, Ulysses B. **Essential concepts about rolling bearings and plain bearings.** Industrial Mechatronics Technology, Garça Technology College (FATEC).

MURPHY, T.; RIENSTRA, A. 2009. **Hear more: a guide to using ultrasound for leak detection and condition monitoring.** Fort Myers, Reliabilityweb.com, 166 p.

SDT 170 MD. 2010 - Available at: http://www.sdthearmore.com/products/sdt170. Accessed on: 24/02/2021.

BRASKEM/TRIUNFO.2007-2010.**Historical of maintenance in industrial**

equipment.

SEQUEIRA, Cláudia. Sensors for measuring mechanical vibrations -

Accelerometers. Available
at:<http://revistamanutencao.pt/PDF/116/M116AT1.pdf>. Accessed on: 02 Feb
2021

JUNIOR, JOE L.R.; **Vibration Analysis in Industrial Bearings,** CENTRO
UNIVERSITÁRIO UNIFACVEST, Lages - SC. Brazil, 2018.

WEBER, Abilio José; et al. Telecurso: professionalising mechanics:
maintenance. Rio de Janeiro: Roberto Marinho Foundation, 2009.

PRUFTECHNIK. JUMPin the VIBXPERT world. Available at:
<https://www.pruftechnik.com/landing-pages/feature-aktion-cm/jumpin-the-
vibxpert- world.html>. Accessed on: 02 May 2021.

PRUFTECHNIK. VIBXPERT II. Available at:
<https://www.pruftechnik.com/br/produtos/sistemas-condition-
monitoring/equipamentos-de- medicao-portateis-para-condition-
monitoring/vibxpert- ii.html>. Accessed on 28 May 2021.

RPM SUL, 2018. Available at: https://rpmsul.com.br/. Accessed on 28 May
2021.

NSK. Bearings. Available at:
<http://nsk.com.br/upload/file/Cat%C3%A1logo%20Geral%20NSK(1).pdf>.
Accessed on: 26 May 2021.

Marcondes. Cast iron irons p. 8, withoutyear. Available at:

<http://ftp.demec.ufpr.br/disciplinas/TM206/Prof_Marcondes/Ferros_fundidos.pd
f>.Accessed on 14 June 2021.

Classification of cast irons. Available at: <
https://www.infomet.com.br/site/acos-e- ligas-conteudo-
ler.php?codConteudo=145>. Accessed on 15 June 2021.

Tensile test. Available at:<https://biopdi.com.br/wp-
content/uploads/2021/04/grafico- tensile-strain-testing-metals-general.jpg>.
Accessed on 16 June 2021.

ANNEXES

Annex A - Classification of radial bearings

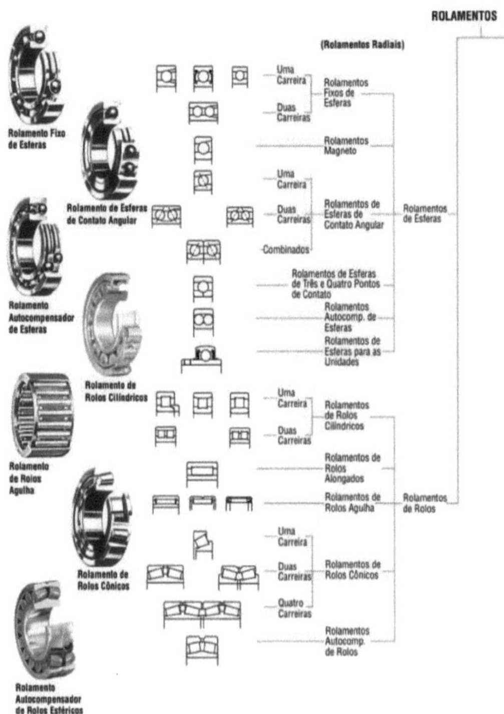

Figure A1 - Classification of radial bearings (Adapted from NSK Brasil, 2013).

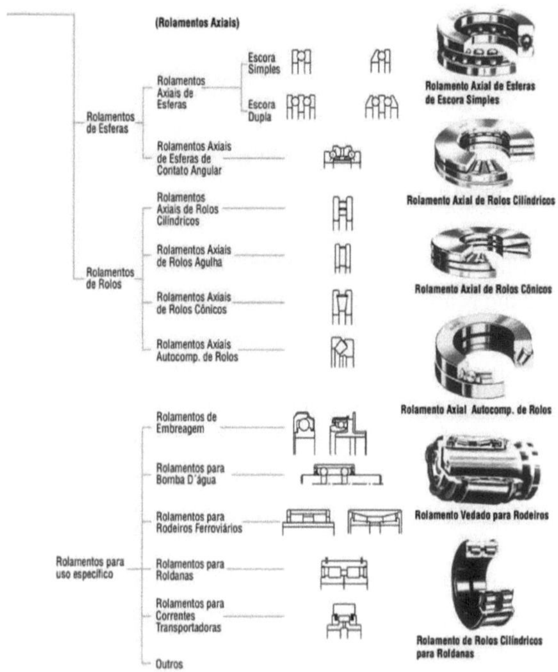

Figure B1 - Classification of thrust bearings (Adapted from NSK Brasil, 2013).

APPENDIX

Appendix A - Dimensioning a bearing

Generally speaking, when sizing a rolling bearing, according to (Silva, Santos and Fernandes, no date, p.7-9), a number of parameters are usually taken into account, including the following:

Stage I: choosing the type of bearing: When choosing the type of bearing, the type of stress to which it will be subjected is taken into account.

Step II: choosing the bearing size: Once you have chosen the type of bearing, the next step is to choose the size of the bearing, using manufacturers' catalogues, based on the outer diameter of the shaft on which the bearing will be mounted, which is called the same as the inner diameter of the bearing.

Stage III: checks: After steps I and II, the bearing capacities are taken from the manufacturer's catalogue, which are:

a) Maximum working speed: nmax is the maximum speed at which the bearing can work for static load (n<10 rpm) and dynamic load (n>=10 rpm).

b) Static load capacity: C0 - is the static load that causes a permanent plastic deformation in the rolling element or track of the order of 1/10000 times the diameter of the rotating element.

c) Dynamic load capacity: C - is the load under which 90 per cent of experienced bearings reach a life of one million revolutions without showing signs of fatigue.

The checks carried out are as follows:

a) Maximum working speed Ntrab$\leq$Nmax

b) Static load capacity C0 min$\leq$C0

With: C0min = S_0 .P0 and P0= X0. Pr + Y0. Pa

Table A1 below gives the S values$_0$ according to the function to be employed by the bearing.

Table A1 - So values (adapted from Silva, Santos and Fernandes, no date, p.8).

Aplicação	s_0
Rolamentos que não giram	
Pás de hélice de passo variável para aviões	$\geq 0,5$
Instalações de comportas de barragens, aliviadores e eclusas	≥ 1
Pontes móveis	$\geq 1,5$
Ganchos de grandes guindastes sem forças dinâmicas adicionais significativas	≥ 1
Ganchos de pequenos guindastes para mercadorias a granel com forças dinâmicas adicionais consideráveis	$\geq 1,6$
Rolamentos em rotação	
onde o serviço é suave e sem vibrações	0,5
onde o serviço e as condições de vibração são normais	1
onde atuam intensas cargas de choque	1,5 a 2

Still for the Coefficient of Safety for Calculating Bearings Static Load Capacity we have:The values of the X_0 and Y_0 coefficients depend on the type of bearing being considered and the ratio between the radial and axial forces on the bearing. For each case, they are given in the manufacturers' catalogues.

c)   Dynamic load capacity: The life of the Lesp $\geq$Ldes bearing

$$\text{Com}: \text{Lesp} = \frac{1000000}{60.n} \left(\frac{C}{P}\right)$$

$$\text{Com}: P = X.\, Pr + Y.Pa$$

With: P = X. Pr + Y.Pa
Where X and Y have the same considerations as X_0 and Y_0 . p is a function of the type of the bearing, and is: 3 for ball bearings 10/3 for roller bearings.

Table A2 below suggests desired life values for bearings, depending on the application.

Table A2 - Desired life values for bearings (Adapted from Silva, Santos and Fernandes, no date, p.9).

Guia para valores de vida nominal requerida L_{10h} em diferentes classes de máquinas

Classe de máquina	L_{10h} horas de trabalho
Eletrodomésticos, máquinas agrícolas, instrumentos, aparelhos técnicos para uso médico	300 a 3 000
Máquinas utilizadas em curtos períodos ou intermitentemente: Máquinas ferramentas manuais, dispositivos de elevação em oficinas, máquinas para construções	3 000 a 8 000
Máquinas para trabalhar com alta confiabilidade durante períodos curtos ou intermitentemente: Elevadores, guindastes para produtos embalados, ou amarras de tambores, fardos, etc.	8 000 a 12 000
Máquinas para 8 horas de trabalho, não totalmente utilizadas: Transmissões de engrenagens para uso geral, motores elétricos para uso industrial, trituradores rotativos, etc.	10 000 a 25 000
Máquinas para 8 horas de trabalho diário, totalmente utilizadas: Máquinas ferramentas, máquinas para trabalhar madeira, máquinas para indústria mecânica em geral, gruas para materiais a granel, ventiladores, correias transportadoras, máquinas de impressão, centrífugas e separadores	20 000 a 30 000
Máquinas para trabalho contínuo, 24 horas por dia: Caixas de pinhões para laminadores, maquinário elétrico de porte médio, compressores, elevadores de minas, bombas, máquinas têxteis	40 000 a 50 000
Equipamentos de abastecimento de água, fornos rotativos, torcedeiras de cabos, máquinas propulsoras de navios	60 000 a 100 000
Máquinas para fabricação de celulose e papel, máquinas elétricas de grande porte, centrais de energia, bombas e ventiladores para minas, mancais de eixos propulsores de navios	~ 100 000

Also according to (Silva, Santos and Fernandes, undated, p.7-9), with regard to the useful life of the bearing to choose the material and adopt the size of the bearing, practical tables are used, such as table A2 above for greater performance of the bearing to be used, and in addition, for the functionality of the bearing to be perfect it is necessary to follow all the manufacturer's recommendations, carry out periodic maintenance and always check for minor damage, so that preventive measures can be taken.

Printed by Books on Demand GmbH, Norderstedt / Germany